LONG DIVISION WORKBOOK

LEARN TO DIVIDE

DOUBLE, TRIPLE, & MULTI-DIGIT

AGES 8+

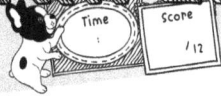

PRACTICE 100 DAYS OF MATH DRILLS

with

RONNY the FRENCHIE

TABLE OF CONTENTS

Answer Keys Included at the Back

INTRODUCTION

Hello there!

It's your favorite fact-sniffing, knowledge-devouring dog, Ronny the Frenchie! I'm very excited to introduce you to this next book because it is all about division! If you haven't yet read or completed one of my other math books, then I'll give you a quick rundown. I started off as a normal French Bulldog, sniffing around for bananas and then devouring them as soon as I found them. My search for delicious bananas took me up the Eiffel Tower one stormy day, and as I reached the top, the tower was struck by lightning! It was amazing that little me was still alive, but also miraculous that my brain had grown bigger, and had a hunger of its own – knowledge! Since then I have been traveling the world and sniffing out facts… and bananas!

In this book, we'll be covering division, which is the opposite of multiplication. Division is the act of separating something into separate parts. This was a tricky one for me because I got used to adding things together and multiplying them, like having 3 bunches of 5 bananas, so 15 bananas. Then I worked out that by doing the opposite, I made division a whole lot easier!

If I have 20 bananas and I need to divide them by 4, I just need to work out how many bananas would need to be in each bunch if I had 4 bunches of bananas. If I had 4 bananas in each bunch, then 4 x 4 would be 16, so that isn't enough, but if I had 5 bananas in each bunch, then 4 x 5 equals 20!

So come along and we can learn more about division together, and along the way, I'll have some fantastic facts for you that will blow your mind! We can practice together and by the end of this book, I'm sure you'll be a division pro!

LONG DIVISION WITH NO REMAINDER

Not sure how to solve the problem? Not to worry! Let's break the problem up into smaller steps.

1)

```
      ┌─────────
  4 ) │ 9   2
      │ ──→
```

2)

```
      │ 2
  4 ) ├─────────
      │ 9   2
      │ 8
```

3)

```
      │ 2
  4 ) ├─────────
      │ 9   2
      │ 8  ↓
      │ ───
      │ 1   2
```

4)

```
      │ 2   3
  4 ) ├─────────
      │ 9   2
      │ 8  ↓
      │ ───
      │ 1   2
      │ 1   2
      │ ───
      │     0
```

1. Start by dividing the digit from left to right.

2. In this problem, we will start by dividing 9 by the divisor, 4. We can ask ourselves, "How many 4s will it take to make 9 or almost 9?" Two 4s would be 8 and that's almost 9, so write down 2 above 9 and 8 below 9.

3. To determine the remainder, subtract 8 from 9. Bring down the number in the next column too and you get 2. The remainder is 12!

4. Now, divide 12 by 4. How many 4s will it take to get 12? The answer is 3! There's no remainder left, and we have the final answer: 23!

NOW TRY IT YOURSELF!

DAY 1

2-Digit by 1-Digit Division

1) 4) 1 2

2) 6) 7 2

3) 4) 4 4

4) 2) 2 6

5) 7) 7 7

6) 8) 8 0

7) 6) 1 2

8) 3) 4 5

9) 7) 1 4

10) 4) 5 2

11) 2) 2 6

12) 2) 4 8

13) 7) 9 1

14) 5) 7 0

15) 2) 5 0

DAY 2

2-Digit by 1-Digit Division

1) 4) 2 8

2) 2) 8 6

3) 7) 7 7

4) 3) 8 7

5) 2) 1 4

6) 4) 7 6

7) 5) 8 0

8) 6) 2 4

9) 3) 9 3

10) 7) 9 8

11) 7) 7 7

12) 5) 7 0

13) 8) 8 0

14) 3) 1 5

15) 7) 9 1

DAY 3

2-Digit by 1-Digit Division

1) 9) 1 8

2) 2) 6 2

3) 9) 9 0

4) 6) 2 4

5) 8) 4 0

6) 3) 8 7

7) 3) 9 3

8) 2) 4 4

9) 2) 8 6

10) 6) 3 6

11) 2) 7 6

12) 3) 8 7

13) 2) 3 4

14) 2) 8 0

15) 5) 6 5

DAY 4

2-Digit by 1-Digit Division

Time :

Score / 15

1) 2)7 8

2) 2)8 4

3) 2)9 4

4) 2)6 2

5) 3)2 1

6) 2)7 4

7) 2)8 6

8) 4)2 4

9) 7)4 9

10) 2)9 4

11) 7)4 9

12) 4)7 6

13) 2)8 6

14) 6)1 2

15) 7)1 4

DAY 5

2-Digit by 1-Digit Division

1) 5) 8 5

2) 2) 7 4

3) 2) 3 8

4) 3) 9 9

5) 5) 7 0

6) 6) 9 6

7) 4) 4 0

8) 5) 2 5

9) 6) 1 8

10) 2) 6 8

11) 2) 1 4

12) 7) 2 8

13) 9) 2 7

14) 3) 6 6

15) 4) 8 0

DAY 6

2-Digit by 1-Digit Division

1) 3) 8 7

2) 5) 9 5

3) 5) 4 5

4) 2) 5 8

5) 5) 5 0

6) 6) 7 2

7) 3) 3 9

8) 2) 4 2

9) 4) 1 6

10) 5) 2 5

11) 3) 2 1

12) 6) 3 0

13) 3) 7 2

14) 4) 2 0

15) 4) 8 8

DAY 7

2-Digit by 1-Digit Division

1) 4) 2 8

2) 8) 8 8

3) 4) 1 2

4) 2) 7 4

5) 2) 6 2

6) 5) 4 0

7) 3) 3 9

8) 2) 4 6

9) 7) 4 9

10) 5) 8 5

11) 2) 5 0

12) 4) 3 2

13) 2) 4 6

14) 5) 2 5

15) 3) 8 7

DAY 8

2-Digit by 1-Digit Division

Time :

Score

/ 15

1) 5) 5 0

2) 4) 4 8

3) 8) 6 4

4) 5) 8 5

5) 2) 7 2

6) 3) 9 0

7) 2) 2 8

8) 7) 3 5

9) 2) 4 6

10) 3) 8 1

11) 3) 8 7

12) 2) 5 4

13) 2) 1 4

14) 2) 1 2

15) 5) 9 0

DAY 9

2-Digit by 1-Digit Division

1) 7) 9 1

2) 3) 9 3

3) 5) 2 5

4) 5) 6 0

5) 2) 1 0

6) 2) 2 6

7) 2) 8 2

8) 2) 7 4

9) 4) 4 4

10) 2) 8 6

11) 3) 9 9

12) 5) 8 5

13) 2) 8 2

14) 3) 2 7

15) 9) 6 3

DAY 10

2-Digit by 1-Digit Division

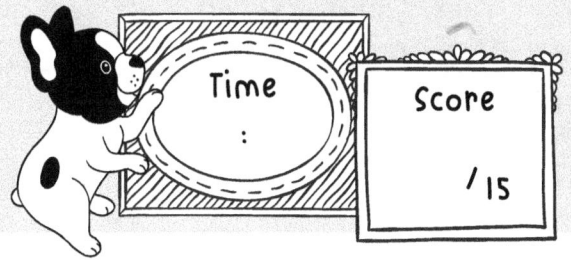

1)
$$2 \overline{)\ 3\ 8\ }$$

2)
$$2 \overline{)\ 8\ 2\ }$$

3)
$$3 \overline{)\ 6\ 3\ }$$

4)
$$2 \overline{)\ 6\ 2\ }$$

5)
$$2 \overline{)\ 7\ 2\ }$$

6)
$$3 \overline{)\ 9\ 3\ }$$

7)
$$4 \overline{)\ 1\ 6\ }$$

8)
$$3 \overline{)\ 9\ 3\ }$$

9)
$$2 \overline{)\ 3\ 8\ }$$

10)
$$5 \overline{)\ 5\ 5\ }$$

11)
$$2 \overline{)\ 5\ 0\ }$$

12)
$$7 \overline{)\ 7\ 0\ }$$

13)
$$2 \overline{)\ 2\ 4\ }$$

14)
$$7 \overline{)\ 5\ 6\ }$$

15)
$$9 \overline{)\ 8\ 1\ }$$

LONG DIVISION WITH REMAINDER

1)

```
        R
4 ) 3 0
```

2)

```
        7 R 2
4 ) 3 0
    2 8
        2
```

Now let's try division problems with remainder:

1. Repeat what we have practiced so far. In this question, we know that it takes seven 4s to get to 28, which is the closest to 30.

2. We are left with the remainder of 2. What should we do? Express your answer as 7 R 2! Easy, right?

DAY 11

2-Digit by 1-Digit Division

1) 9) 1 5 R

2) 3) 9 4 R

3) 4) 2 8 R

4) 8) 8 7 R

5) 3) 7 1 R

6) 4) 6 3 R

7) 8) 9 0 R

8) 4) 4 2 R

9) 6) 8 2 R

10) 2) 3 9 R

11) 7) 6 8 R

12) 9) 1 1 R

13) 4) 5 6 R

14) 3) 2 3 R

15) 8) 5 3 R

DAY 12

2-Digit by 1-Digit Division

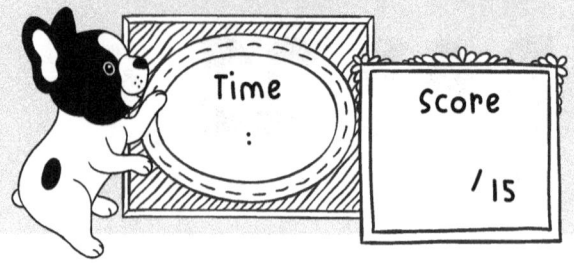

1) 5) 7 9 R

2) 6) 1 1 R

3) 9) 7 7 R

4) 9) 7 8 R

5) 2) 8 2 R

6) 5) 6 2 R

7) 6) 1 5 R

8) 3) 6 1 R

9) 2) 8 3 R

10) 9) 7 2 R

11) 5) 1 0 R

12) 6) 2 1 R

13) 3) 2 3 R

14) 7) 8 3 R

15) 9) 8 4 R

DAY 13

2-Digit by 1-Digit Division

1) 3) 9 1   R

2) 6) 4 2   R

3) 2) 2 5   R

4) 3) 4 7   R

5) 4) 7 5   R

6) 8) 7 3   R

7) 3) 9 4   R

8) 5) 7 1   R

9) 6) 4 9   R

10) 8) 8 8   R

11) 5) 2 5   R

12) 2) 2 2   R

13) 8) 5 2   R

14) 5) 4 8   R

15) 8) 4 2   R

DAY 14

2-Digit by 1-Digit Division

Time :

Score / 15

1) 6) 8 2 R

2) 3) 1 9 R

3) 5) 9 1 R

4) 5) 4 5 R

5) 8) 8 8 R

6) 2) 6 9 R

7) 8) 6 5 R

8) 4) 4 8 R

9) 4) 9 6 R

10) 4) 8 8 R

11) 7) 8 3 R

12) 9) 7 8 R

13) 3) 6 4 R

14) 7) 6 7 R

15) 6) 9 1 R

DAY 15

2-Digit by 1-Digit Division

1) 2) 4 7 R

2) 7) 9 1 R

3) 4) 9 1 R

4) 7) 2 7 R

5) 3) 6 9 R

6) 9) 8 4 R

7) 9) 1 1 R

8) 4) 2 0 R

9) 2) 5 3 R

10) 9) 3 8 R

11) 8) 5 5 R

12) 9) 3 9 R

13) 3) 2 8 R

14) 8) 9 3 R

15) 9) 3 2 R

DAY 16

2-Digit by 1-Digit Division

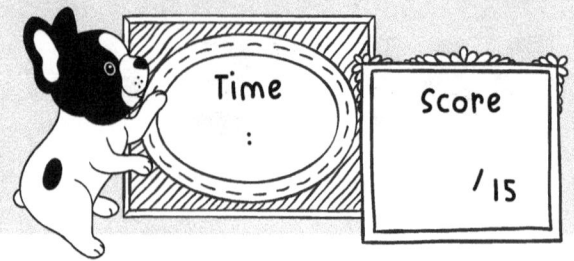

1) $8 \overline{)\ 1 \quad 9}$ R

2) $6 \overline{)\ 4 \quad 5}$ R

3) $9 \overline{)\ 8 \quad 9}$ R

4) $9 \overline{)\ 9 \quad 0}$ R

5) $5 \overline{)\ 2 \quad 1}$ R

6) $6 \overline{)\ 6 \quad 0}$ R

7) $5 \overline{)\ 8 \quad 2}$ R

8) $3 \overline{)\ 8 \quad 6}$ R

9) $5 \overline{)\ 7 \quad 9}$ R

10) $6 \overline{)\ 9 \quad 0}$ R

11) $3 \overline{)\ 6 \quad 8}$ R

12) $5 \overline{)\ 6 \quad 6}$ R

13) $6 \overline{)\ 1 \quad 4}$ R

14) $5 \overline{)\ 7 \quad 3}$ R

15) $5 \overline{)\ 8 \quad 5}$ R

DAY 17

2-Digit by 1-Digit Division

1) 6) 7 9 R

2) 5) 1 5 R

3) 3) 2 4 R

4) 9) 6 4 R

5) 5) 8 7 R

6) 6) 4 8 R

7) 2) 4 3 R

8) 7) 1 0 R

9) 4) 4 7 R

10) 3) 7 7 R

11) 6) 7 5 R

12) 8) 1 9 R

13) 3) 7 7 R

14) 5) 3 1 R

15) 6) 9 6 R

DAY 18

2-Digit by 1-Digit Division

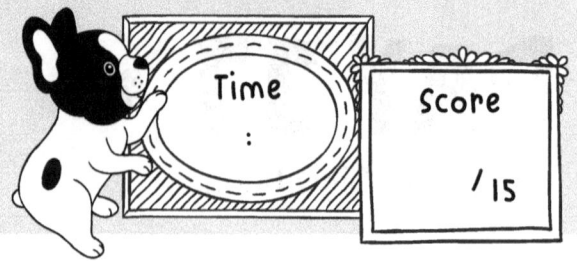

1) 5) 6 7 R

2) 6) 9 5 R

3) 4) 7 7 R

4) 2) 4 5 R

5) 8) 8 0 R

6) 9) 3 0 R

7) 8) 7 0 R

8) 5) 8 5 R

9) 8) 9 5 R

10) 2) 1 5 R

11) 9) 7 4 R

12) 2) 7 5 R

13) 8) 5 4 R

14) 3) 1 0 R

15) 4) 9 8 R

DAY 19

2-Digit by 1-Digit Division

1) 7) 9 7 R

2) 2) 3 1 R

3) 4) 5 5 R

4) 6) 7 6 R

5) 5) 8 4 R

6) 6) 6 7 R

7) 6) 6 5 R

8) 7) 7 1 R

9) 5) 5 0 R

10) 3) 7 4 R

11) 4) 3 7 R

12) 2) 6 3 R

13) 7) 5 4 R

14) 9) 4 9 R

15) 2) 4 7 R

DAY 20

2-Digit by 1-Digit Division

1) 9) 2 5   R

2) 9) 6 8   R

3) 5) 9 6   R

4) 6) 7 3   R

5) 7) 6 9   R

6) 9) 9 1   R

7) 8) 3 7   R

8) 8) 5 4   R

9) 4) 4 5   R

10) 5) 2 8   R

11) 7) 2 5   R

12) 3) 8 0   R

13) 2) 3 5   R

14) 9) 9 1   R

15) 3) 3 8   R

WHAT IS THE SMALLEST NUMBER DIVISIBLE BY 1, 2, 3, 4, 5, 6, 7, 8, 9, AND 10?

2520, that is a lot of bananas in bunches! What would I do if I had 2520 bananas? I think I would start by eating as many bananas as I could, and then make as many banana-related foods as I can think of. Banana bread, banana pancakes, banana smoothies, frozen bananas covered in chocolate, you name it! What would you do if you had 2520 bananas?

DAY 21

3-Digit by 1-Digit Division

Time :

Score / 15

1) 6 ⟌ 9 2 4

2) 7 ⟌ 3 7 1

3) 5 ⟌ 2 0 5

4) 5 ⟌ 1 3 0

5) 5 ⟌ 2 1 5

6) 3 ⟌ 5 9 7

7) 3 ⟌ 3 2 7

8) 2 ⟌ 2 6 4

9) 7 ⟌ 3 7 1

10) 6 ⟌ 7 9 8

11) 3 ⟌ 7 6 5

12) 3 ⟌ 6 8 1

13) 5 ⟌ 7 0 5

14) 2 ⟌ 2 5 6

15) 3 ⟌ 4 2 3

DAY 22

3-Digit by 1-Digit Division

1) 5) 9 8 5

2) 5) 9 2 5

3) 2) 7 3 8

4) 2) 7 2 4

5) 9) 8 3 7

6) 5) 7 9 5

7) 4) 6 6 8

8) 3) 7 1 7

9) 2) 3 8 6

10) 3) 2 2 2

11) 8) 3 0 4

12) 7) 8 2 6

13) 4) 9 2 8

14) 9) 8 1 9

15) 5) 9 1 0

DAY 23

3-Digit by 1-Digit Division

1) $9 \overline{)963}$

2) $2 \overline{)534}$

3) $5 \overline{)955}$

4) $5 \overline{)155}$

5) $4 \overline{)304}$

6) $2 \overline{)894}$

7) $2 \overline{)106}$

8) $2 \overline{)172}$

9) $2 \overline{)398}$

10) $2 \overline{)942}$

11) $3 \overline{)618}$

12) $5 \overline{)565}$

13) $9 \overline{)252}$

14) $2 \overline{)896}$

15) $5 \overline{)975}$

DAY 24

3-Digit by 1-Digit Division

1) 3) 9 9 9

2) 6) 8 1 6

3) 2) 5 6 6

4) 3) 5 9 1

5) 4) 8 4 4

6) 4) 4 5 2

7) 9) 2 3 4

8) 2) 8 1 8

9) 7) 9 9 4

10) 6) 6 9 0

11) 7) 6 7 9

12) 6) 8 2 2

13) 7) 6 2 3

14) 2) 5 0 6

15) 4) 6 3 2

DAY 25

3-Digit by 1-Digit Division

1) 2) 1 3 0

2) 2) 9 4 8

3) 2) 2 2 2

4) 6) 4 9 8

5) 3) 8 4 9

6) 2) 4 4 6

7) 3) 8 6 4

8) 7) 4 8 3

9) 2) 2 6 0

10) 5) 7 4 0

11) 9) 8 4 6

12) 3) 2 7 6

13) 2) 4 5 2

14) 5) 6 8 5

15) 4) 6 0 8

DAY 26

3-Digit by 1-Digit Division

1) 3) 2 1 9

2) 3) 4 5 3

3) 2) 8 1 4

4) 5) 6 5 5

5) 4) 3 0 4

6) 4) 2 8 4

7) 2) 7 8 8

8) 5) 7 7 0

9) 3) 7 2 3

10) 2) 5 1 4

11) 7) 8 7 5

12) 2) 5 5 6

13) 4) 3 9 2

14) 4) 7 1 6

15) 3) 8 4 9

DAY 27

3-Digit by 1-Digit Division

Time
:

Score
/ 15

1) 2 ⟌ 7 7 8

2) 3 ⟌ 6 6 3

3) 2 ⟌ 2 6 0

4) 5 ⟌ 7 4 0

5) 5 ⟌ 8 5 5

6) 4 ⟌ 4 1 6

7) 3 ⟌ 3 2 7

8) 5 ⟌ 6 1 5

9) 2 ⟌ 2 9 0

10) 3 ⟌ 6 3 3

11) 4 ⟌ 4 5 2

12) 8 ⟌ 8 8 0

13) 2 ⟌ 6 3 4

14) 7 ⟌ 5 9 5

15) 2 ⟌ 7 3 4

DAY 28 🐾

3-Digit by 1-Digit Division

1) 4) 4 0 4

2) 5) 6 1 0

3) 2) 4 3 0

4) 3) 8 4 3

5) 8) 6 1 6

6) 5) 2 2 0

7) 3) 6 8 4

8) 4) 3 1 6

9) 2) 9 8 2

10) 9) 5 1 3

11) 6) 9 6 6

12) 3) 8 4 3

13) 5) 2 1 0

14) 5) 5 3 5

15) 3) 3 2 1

DAY 29

3-Digit by 1-Digit Division

1) 3) 5 4 3

2) 2) 8 1 6

3) 6) 7 9 2

4) 2) 3 1 6

5) 3) 3 9 3

6) 7) 5 5 3

7) 2) 9 9 8

8) 3) 5 0 1

9) 5) 3 5 5

10) 3) 9 0 6

11) 2) 4 0 2

12) 2) 3 7 6

13) 2) 9 5 0

14) 3) 1 8 6

15) 2) 8 8 8

DAY 30

3-Digit by 1-Digit Division

1) 2) 7 8 2

2) 4) 8 2 4

3) 3) 1 2 3

4) 2) 9 8 2

5) 3) 2 6 7

6) 5) 2 4 5

7) 2) 8 9 0

8) 2) 4 3 6

9) 7) 2 0 3

10) 3) 6 4 2

11) 2) 7 2 2

12) 5) 5 4 5

13) 9) 9 9 9

14) 2) 4 3 6

15) 2) 6 9 4

YOU CAN CUT A CIRCULAR CAKE INTO 8 EQUAL PIECES WITH ONLY 3 CUTS. DO YOU KNOW HOW?

Try it out first and then read the answer.

There's a bit of a trick to this one, and I did not think about it the first time a pastry chef gave me this challenge because the cake smelled so delicious!

First, you cut the cake in half one way. Second, cut the cake in half the other way, giving you 4 equal pieces. Now here's the trick: instead of cutting down into the cake again, cut into the side instead, through the middle of all 4 pieces, and voila! 8 pieces of delicious cake.

DAY 31

3-Digit by 1-Digit Division

1) 7) 6 7 2   R

2) 4) 1 5 6   R

3) 4) 9 4 9   R

4) 3) 3 1 7   R

5) 2) 1 4 5   R

6) 9) 3 5 1   R

7) 3) 4 9 9   R

8) 8) 3 5 4   R

9) 2) 1 9 1   R

10) 4) 6 0 4   R

11) 4) 8 2 5   R

12) 3) 1 0 6   R

DAY 32

3-Digit by 1-Digit Division

Score

/ 12

1) 4) 2 0 9   R

2) 8) 1 0 7   R

3) 6) 1 4 4   R

4) 9) 9 7 9   R

5) 9) 9 4 6   R

6) 8) 2 9 1   R

7) 5) 2 1 3   R

8) 6) 2 4 9   R

9) 6) 4 1 1   R

10) 7) 1 8 7   R

11) 8) 1 6 2   R

12) 4) 4 1 1   R

DAY 33

3-Digit by 1-Digit Division

1) 7) 9 2 3 R

2) 8) 1 3 4 R

3) 7) 5 4 7 R

4) 7) 5 4 2 R

5) 8) 2 8 9 R

6) 4) 5 1 4 R

7) 5) 8 9 1 R

8) 8) 6 7 9 R

9) 8) 5 1 3 R

10) 7) 4 6 9 R

11) 6) 9 4 5 R

12) 2) 2 0 6 R

DAY 34

3-Digit by 1-Digit Division

1) 3) 3 8 8 R

2) 6) 3 3 0 R

3) 8) 6 6 1 R

4) 2) 5 5 2 R

5) 5) 7 6 4 R

6) 2) 3 0 4 R

7) 2) 8 0 0 R

8) 4) 7 1 4 R

9) 6) 2 3 6 R

10) 7) 5 5 8 R

11) 3) 4 2 2 R

12) 8) 1 4 8 R

DAY 35

3-Digit by 1-Digit Division

1) 7) 2 8 0   R

2) 2) 9 2 8   R

3) 8) 1 9 6   R

4) 3) 1 8 8   R

5) 2) 2 8 0   R

6) 3) 1 6 4   R

7) 4) 7 4 2   R

8) 5) 6 0 9   R

9) 4) 7 7 1   R

10) 9) 6 2 2   R

11) 7) 1 2 5   R

12) 7) 9 9 7   R

DAY 36

3-Digit by 1-Digit Division

1)

6) 6 2 1   R

2)

7) 7 5 7   R

3)

6) 8 8 6   R

4)

6) 4 1 7   R

5)

8) 3 7 9   R

6)

6) 5 8 0   R

7)

7) 7 1 4   R

8)

5) 6 2 5   R

9)

6) 4 7 2   R

10)

2) 3 4 7   R

11)

5) 2 3 2   R

12)

6) 1 7 6   R

DAY 37

3-Digit by 1-Digit Division

1) 6)176 R

2) 9)101 R

3) 3)908 R

4) 3)692 R

5) 8)489 R

6) 2)358 R

7) 6)493 R

8) 9)132 R

9) 7)993 R

10) 6)188 R

11) 3)758 R

12) 6)193 R

DAY 38

3-Digit by 1-Digit Division

1) 4)202 R

2) 2)950 R

3) 6)484 R

4) 5)824 R

5) 5)978 R

6) 6)159 R

7) 6)301 R

8) 9)787 R

9) 2)107 R

10) 4)292 R

11) 2)848 R

12) 3)369 R

DAY 39

3-Digit by 1-Digit Division

Time
:

Score

/ 12

1) 5) 7 5 4 R

2) 5) 1 1 7 R

3) 2) 7 6 1 R

4) 9) 1 7 3 R

5) 9) 1 7 2 R

6) 2) 8 1 2 R

7) 8) 2 0 1 R

8) 4) 4 4 6 R

9) 4) 6 5 8 R

10) 4) 7 8 8 R

11) 3) 2 3 7 R

12) 8) 8 4 6 R

DAY 40

3-Digit by 1-Digit Division

1) 9) 9 6 9   R

2) 9) 4 5 4   R

3) 3) 3 9 5   R

4) 6) 2 1 1   R

5) 9) 4 4 0   R

6) 2) 5 4 2   R

7) 9) 5 3 6   R

8) 5) 2 7 7   R

9) 9) 2 2 0   R

10) 3) 1 1 7   R

11) 3) 2 1 0   R

12) 7) 9 7 6   R

LET'S GO BACK IN TIME FOR THIS FACT!

Thousands of years ago, Egypt was one of the most powerful and prosperous civilizations in the world. One of the greatest achievements they are still known for to this day is the Great Pyramids of Giza. The largest of these pyramids stood at about 138 meters tall, that's about 41 stories!

Not only are these pyramids one of the great wonders of the world, they are also a mathematical wonder! According to historians and scholars, the design embodies the concepts of the Pythagorean Theorem, the Golden Ratio, and Pi. Despite its amazing mathematical and historical importance, we still don't know everything there is to know about the Pyramids. I sure hope we find out more, I'm just itching to know!

DAY 41

4-Digit by 1-Digit Division

1) 5) 3 4 5 5

2) 3) 6 3 9 3

3) 8) 1 2 2 4

4) 2) 8 7 7 6

5) 7) 2 6 5 3

6) 3) 7 2 5 7

7) 3) 4 8 9 9

8) 5) 8 8 2 5

9) 5) 4 8 1 5

10) 2) 7 6 2 2

11) 4) 7 0 7 2

12) 5) 6 4 1 5

DAY 42 🐾

4-Digit by 1-Digit Division

1) 5) 1 1 7 5

2) 3) 4 6 0 2

3) 2) 1 3 0 6

4) 7) 5 5 3 7

5) 4) 9 3 3 2

6) 2) 4 0 0 4

7) 3) 1 3 5 9

8) 6) 2 5 3 8

9) 6) 4 4 3 4

10) 2) 2 9 1 6

11) 2) 5 6 7 2

12) 3) 2 3 2 5

DAY 43

4-Digit by 1-Digit Division

1) 7) 9 7 5 8

2) 2) 1 5 0 2

3) 3) 9 7 8 0

4) 4) 9 5 9 6

5) 3) 9 3 2 1

6) 2) 7 6 5 8

7) 2) 8 2 1 2

8) 5) 8 5 4 5

9) 2) 8 9 4 2

10) 4) 2 3 5 6

11) 9) 1 7 5 5

12) 2) 2 5 7 8

DAY 44

4-Digit by 1-Digit Division

1) 2) 2 3 7 6

2) 5) 9 3 3 5

3) 8) 8 9 2 8

4) 5) 4 2 6 5

5) 8) 9 2 6 4

6) 5) 3 7 8 0

7) 5) 3 6 5 5

8) 8) 6 7 2 0

9) 3) 4 6 2 9

10) 5) 8 0 5 5

11) 2) 2 9 7 2

12) 7) 3 0 3 1

DAY 45

4-Digit by 1-Digit Division

Time :

Score / 12

1) 6) 1 9 0 2

2) 7) 4 0 2 5

3) 2) 2 0 2 4

4) 4) 5 3 3 6

5) 6) 8 2 9 2

6) 2) 7 5 5 8

7) 7) 7 7 2 1

8) 2) 3 5 0 2

9) 7) 6 3 7 7

10) 7) 4 4 7 3

11) 5) 3 3 1 5

12) 2) 9 1 9 4

DAY 46

4-Digit by 1-Digit Division

1) 6) 2 5 5 6

2) 2) 8 2 0 0

3) 3) 7 6 8 9

4) 4) 4 3 4 8

5) 7) 6 4 6 8

6) 6) 9 4 8 6

7) 2) 3 6 8 6

8) 3) 6 8 3 7

9) 6) 1 7 6 4

10) 4) 4 5 4 8

11) 2) 2 9 2 4

12) 5) 8 3 0 0

DAY 47

4-Digit by 1-Digit Division

1) $3 \overline{)1\ 3\ 4\ 7}$

2) $2 \overline{)7\ 2\ 6\ 8}$

3) $5 \overline{)4\ 8\ 3\ 5}$

4) $7 \overline{)1\ 3\ 3\ 7}$

5) $6 \overline{)9\ 5\ 2\ 8}$

6) $2 \overline{)7\ 5\ 6\ 2}$

7) $7 \overline{)1\ 5\ 4\ 7}$

8) $7 \overline{)8\ 9\ 4\ 6}$

9) $9 \overline{)5\ 6\ 4\ 3}$

10) $7 \overline{)1\ 0\ 8\ 5}$

11) $3 \overline{)3\ 3\ 2\ 7}$

12) $8 \overline{)5\ 6\ 8\ 0}$

DAY 48

4-Digit by 1-Digit Division

Time :

Score

/12

1) 2) 4 5 3 4

2) 5) 8 5 2 5

3) 2) 8 3 8 0

4) 5) 7 9 9 0

5) 5) 8 2 4 5

6) 8) 2 6 6 4

7) 2) 3 9 3 8

8) 5) 7 9 7 5

9) 2) 9 2 2 2

10) 5) 2 5 4 5

11) 2) 9 2 1 4

12) 2) 2 6 8 6

DAY 49

4-Digit by 1-Digit Division

1) 4) 9 4 4 4

2) 2) 4 3 7 2

3) 7) 6 7 9 7

4) 4) 5 6 8 4

5) 7) 6 5 1 7

6) 3) 9 8 9 7

7) 9) 8 1 0 9

8) 2) 9 3 0 2

9) 2) 6 6 1 4

10) 7) 8 2 6 0

11) 5) 7 2 6 5

12) 9) 3 2 4 9

DAY 50 🐾

4-Digit by 1-Digit Division

Time
:

Score

/ 12

1) 2) 9 2 4 2

2) 9) 9 2 4 3

3) 5) 9 9 2 5

4) 2) 4 6 1 8

5) 5) 6 7 6 5

6) 2) 3 7 6 8

7) 2) 3 2 3 8

8) 7) 3 3 2 5

9) 7) 6 6 4 3

10) 6) 4 2 4 8

11) 5) 9 2 9 5

12) 7) 3 5 7 7

DO YOU KNOW WHAT THE MATHEMATICAL NAME FOR THE DIVISION BAR IN A FRACTION IS?

Vinculum!

What a strange name, vinculum.
It sounds like a magic spell.

DAY 51

4-Digit by 1-Digit Division

Time :

Score / 12

1) 5) 6 9 4 9   R

2) 2) 4 4 2 6   R

3) 4) 7 3 5 5   R

4) 6) 9 1 3 3   R

5) 5) 4 7 4 0   R

6) 7) 8 8 2 3   R

7) 4) 1 0 6 6   R

8) 9) 9 5 1 1   R

9) 6) 1 0 2 4   R

10) 6) 7 0 0 7   R

11) 4) 8 4 6 1   R

12) 9) 7 3 9 5   R

DAY 52

4-Digit by 1-Digit Division

1) 3) 9 2 3 2   R

2) 7) 2 9 1 2   R

3) 5) 9 7 1 4   R

4) 9) 5 2 8 1   R

5) 2) 5 0 2 2   R

6) 4) 1 1 6 9   R

7) 5) 6 7 8 3   R

8) 7) 4 4 3 4   R

9) 6) 9 7 2 4   R

10) 9) 4 3 9 2   R

11) 2) 7 1 5 4   R

12) 5) 5 4 4 2   R

DAY 53

4-Digit by 1-Digit Division

1)
8) 4 6 1 3   R

2)
9) 8 1 7 8   R

3)
2) 7 5 1 8   R

4)
6) 6 6 8 7   R

5)
4) 3 8 7 4   R

6)
8) 2 3 3 9   R

7)
9) 6 8 0 5   R

8)
2) 2 3 7 4   R

9)
3) 4 8 9 0   R

10)
8) 1 6 8 5   R

11)
4) 4 5 8 1   R

12)
3) 5 4 6 5   R

DAY 54

4-Digit by 1-Digit Division

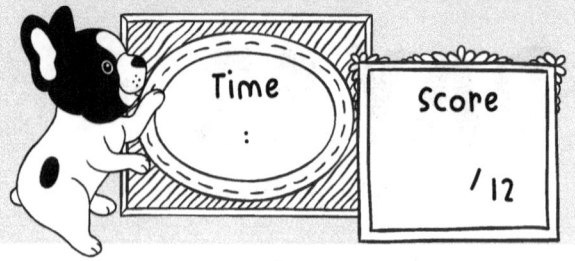

1) 3)5239 R

2) 3)9886 R

3) 7)3817 R

4) 5)2856 R

5) 3)4957 R

6) 9)2194 R

7) 4)4597 R

8) 2)1998 R

9) 4)5744 R

10) 2)3949 R

11) 3)1121 R

12) 9)3800 R

DAY 55

4-Digit by 1-Digit Division

1) 2) 9 1 5 8 R

2) 9) 1 5 7 5 R

3) 3) 1 3 9 3 R

4) 9) 6 2 8 1 R

5) 8) 6 1 9 8 R

6) 2) 2 6 5 8 R

7) 8) 5 5 5 1 R

8) 4) 1 0 0 8 R

9) 4) 6 9 9 5 R

10) 6) 1 4 6 4 R

11) 7) 8 4 1 5 R

12) 3) 3 9 1 2 R

DAY 56

4-Digit by 1-Digit Division

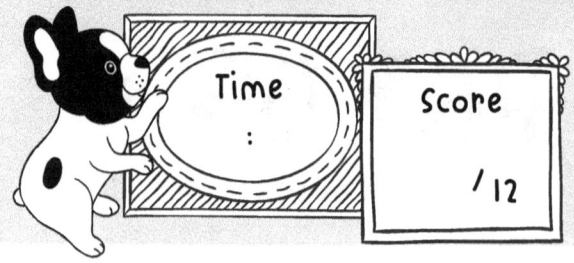

1) 6)‾4‾9‾6‾1 R

2) 9)‾3‾9‾7‾3 R

3) 4)‾4‾5‾7‾9 R

4) 5)‾5‾4‾6‾9 R

5) 9)‾5‾9‾8‾9 R

6) 9)‾1‾8‾2‾5 R

7) 2)‾7‾9‾4‾9 R

8) 9)‾6‾4‾6‾2 R

9) 8)‾7‾5‾7‾5 R

10) 4)‾2‾4‾8‾5 R

11) 9)‾5‾0‾7‾5 R

12) 6)‾5‾0‾1‾4 R

DAY 57

4-Digit by 1-Digit Division

1) 4) 2 9 9 9   R

2) 2) 6 1 4 3   R

3) 8) 7 4 8 3   R

4) 5) 2 9 0 4   R

5) 4) 9 9 4 7   R

6) 5) 5 0 2 3   R

7) 7) 8 4 9 2   R

8) 4) 2 9 5 9   R

9) 5) 5 9 1 1   R

10) 6) 9 5 5 1   R

11) 7) 6 1 8 2   R

12) 4) 8 1 7 7   R

DAY 58

4-Digit by 1-Digit Division

Time :

Score

/ 12

1) 8) 4 2 2 7   R

2) 9) 5 9 5 6   R

3) 5) 9 2 8 1   R

4) 2) 7 3 4 4   R

5) 9) 4 9 9 5   R

6) 5) 7 4 0 4   R

7) 6) 8 3 8 0   R

8) 4) 9 0 5 9   R

9) 3) 3 7 4 6   R

10) 6) 6 5 1 4   R

11) 8) 7 3 8 8   R

12) 6) 3 5 3 2   R

DAY 59

4-Digit by 1-Digit Division

1) 2) 3 4 0 3   R

2) 2) 6 2 9 4   R

3) 5) 8 6 1 8   R

4) 9) 6 0 9 0   R

5) 8) 2 9 4 5   R

6) 2) 6 4 2 2   R

7) 7) 8 9 6 0   R

8) 7) 4 5 4 5   R

9) 7) 3 7 1 4   R

10) 6) 7 6 7 3   R

11) 7) 4 4 6 2   R

12) 4) 8 5 8 1   R

DAY 60 🐾

4-Digit by 1-Digit Division

1)   4) 4 8 3 9   R

2)   6) 9 5 0 1   R

3)   6) 9 5 4 4   R

4)   3) 1 6 9 1   R

5)   3) 4 4 7 7   R

6)   7) 6 9 9 5   R

7)   7) 3 7 7 8   R

8)   9) 8 8 2 6   R

9)   8) 2 3 0 3   R

10)   7) 4 7 9 8   R

11)   8) 2 3 4 8   R

12)   2) 1 1 2 9   R

DID YOU KNOW THAT YOU CANNOT DIVIDE A NUMBER BY 0?

When we divide one number by another, we are asking how many times that number can be subtracted from the other to get to 0.

If we divide by 0, we get an undefined result because we can subtract 0 from our number an infinite amount of times. The issue also occurs when we try and reverse the process.

If we divided 12 by 3, we would get an answer of 4. If we multiply 4 by 3, we get back to 12. This also cannot be done with 0, as multiplying any number by 0 equals 0, which is a paradox. Thus, dividing by 0 is not allowed in math.

I've never been able to understand paradoxes myself, like 'what came first, the banana or the banana tree?'

DAY 61

5-Digit by 1-Digit Division

1)
4) 7 6 6 0 6   R

2)
8) 2 5 3 9 1   R

3)
3) 5 4 4 1 5   R

4)
9) 9 8 3 2 1   R

5)
4) 1 0 3 1 0   R

6)
6) 9 2 4 0 3   R

7)
8) 1 5 3 8 1   R

8)
3) 4 6 5 1 6   R

DAY 62

5-Digit by 1-Digit Division

1) 5) 2 0 8 9 3   R

2) 6) 4 6 6 6 6   R

3) 8) 5 7 3 1 1   R

4) 6) 3 3 9 1 9   R

5) 7) 8 2 1 9 1   R

6) 2) 3 7 5 3 9   R

7) 8) 4 5 0 0 7   R

8) 3) 8 0 5 7 1   R

DAY 63

5-Digit by 1-Digit Division

1) $6 \overline{)41498}$ R

2) $9 \overline{)95013}$ R

3) $7 \overline{)42464}$ R

4) $9 \overline{)33808}$ R

5) $4 \overline{)10385}$ R

6) $2 \overline{)52889}$ R

7) $6 \overline{)61482}$ R

8) $7 \overline{)82283}$ R

DAY 64

5-Digit by 1-Digit Division

1)
$$4 \overline{)1\ 9\ 2\ 8\ 8} \quad R$$

2)
$$4 \overline{)8\ 8\ 8\ 4\ 6} \quad R$$

3)
$$3 \overline{)6\ 4\ 4\ 6\ 2} \quad R$$

4)
$$3 \overline{)8\ 6\ 9\ 9\ 5} \quad R$$

5)
$$5 \overline{)1\ 8\ 7\ 6\ 1} \quad R$$

6)
$$3 \overline{)3\ 7\ 6\ 3\ 2} \quad R$$

7)
$$8 \overline{)2\ 8\ 5\ 4\ 5} \quad R$$

8)
$$3 \overline{)4\ 6\ 3\ 5\ 5} \quad R$$

DAY 65

5-Digit by 1-Digit Division

1)

7) 1 7 3 7 2   R

2)

8) 4 4 0 5 2   R

3)

5) 4 3 3 4 7   R

4)

3) 7 5 7 6 5   R

5)

5) 4 0 8 9 9   R

6)

5) 4 4 8 6 9   R

7)

9) 2 6 0 1 3   R

8)

3) 5 3 0 5 6   R

DAY 66

5-Digit by 1-Digit Division

1) 2) 3 9 7 7 2   R

2) 9) 4 3 6 3 4   R

3) 4) 9 4 1 7 4   R

4) 5) 8 6 8 8 2   R

5) 7) 1 2 1 7 0   R

6) 8) 3 6 7 4 9   R

7) 8) 5 1 1 6 2   R

8) 5) 1 6 3 4 7   R

DAY 67

5-Digit by 1-Digit Division

1) 7) 3 3 4 4 9   R

2) 6) 1 7 8 3 5   R

3) 5) 4 0 5 1 8   R

4) 7) 4 4 3 8 1   R

5) 7) 6 6 8 1 0   R

6) 4) 4 0 0 0 6   R

7) 7) 7 6 5 1 6   R

8) 4) 4 7 7 3 8   R

DAY 68

5-Digit by 1-Digit Division

1) 7) 9 4 1 1 7   R

2) 9) 9 1 4 4 5   R

3) 9) 8 1 4 6 6   R

4) 6) 6 8 9 2 4   R

5) 9) 2 2 5 5 0   R

6) 5) 3 8 5 4 3   R

7) 7) 8 4 6 8 7   R

8) 5) 5 2 6 0 7   R

LONG DIVISION WITH DOUBLE DIGIT DIVISOR

1)

```
         _____
2 2 ) 7 4 8
```

2)

```
           3
         _____
2 2 ) 7 4 8
      6 6
```

3)

```
           3
         _____
2 2 ) 7 4 8
      6 6 ↓
         8 8
```

4)

```
           3 4
         _____
2 2 ) 7 4 8
      6 6 ↓
         8 8
         8 8
           0
```

1.Does this question look difficult to solve? Don't worry. You probably already know how to solve it without realizing it. Take the same approach we did with the single divisor. This time we ask ourselves "How many 22s does it take to fit in 7?" The answer is obvious! It simply would not fit!

2.Let's move from left to right. We ask ourselves again. "How many 22s does it take to fit in 74?" Let's take a guess: 4? Oh no, 22 x 4 = 88. The number is too big. Take another guess: 3? The answer fits into 74 this time because 22 x 3 = 66.

3.What are we left with? 74 - 66 = 8. Bring down the last number, which is 8 again. The remainder is 88.

4.How many 22s fit in 88? You probably know the answer already! The answer is 4. Yay! We have the final answer: 34.

DAY 69

3-Digit by 2-Digit Division

Time :

Score / 15

1) 88) 3 5 2

2) 18) 9 5 4

3) 35) 2 1 0

4) 79) 6 3 2

5) 91) 6 3 7

6) 78) 6 2 4

7) 23) 2 0 7

8) 44) 2 2 0

9) 69) 6 9 0

10) 41) 2 8 7

11) 29) 7 2 5

12) 13) 1 9 5

13) 74) 9 6 2

14) 79) 3 9 5

15) 53) 6 8 9

DAY 70

3-Digit by 2-Digit Division

1) $9\ 3 \overline{)\ 9\ 3\ 0\ }$

2) $5\ 3 \overline{)\ 7\ 9\ 5\ }$

3) $3\ 8 \overline{)\ 6\ 8\ 4\ }$

4) $4\ 5 \overline{)\ 2\ 2\ 5\ }$

5) $8\ 9 \overline{)\ 4\ 4\ 5\ }$

6) $2\ 9 \overline{)\ 5\ 2\ 2\ }$

7) $2\ 5 \overline{)\ 8\ 2\ 5\ }$

8) $2\ 8 \overline{)\ 1\ 9\ 6\ }$

9) $6\ 9 \overline{)\ 2\ 0\ 7\ }$

10) $5\ 4 \overline{)\ 3\ 7\ 8\ }$

11) $4\ 1 \overline{)\ 9\ 4\ 3\ }$

12) $9\ 7 \overline{)\ 4\ 8\ 5\ }$

13) $7\ 9 \overline{)\ 9\ 4\ 8\ }$

14) $3\ 1 \overline{)\ 2\ 1\ 7\ }$

15) $7\ 6 \overline{)\ 3\ 0\ 4\ }$

DAY 71

3-Digit by 2-Digit Division

1) 3 1) 3 4 1

2) 4 8) 1 9 2

3) 3 6) 9 3 6

4) 5 4) 5 9 4

5) 1 5) 3 9 0

6) 1 4) 2 9 4

7) 6 7) 4 0 2

8) 7 9) 3 1 6

9) 1 4) 1 5 4

10) 4 3) 4 7 3

11) 3 9) 8 5 8

12) 5 1) 3 5 7

13) 3 1) 6 5 1

14) 2 1) 9 4 5

15) 9 4) 6 5 8

DAY 72

3-Digit by 2-Digit Division

1) $11\overline{)913}$

2) $15\overline{)165}$

3) $37\overline{)333}$

4) $86\overline{)172}$

5) $24\overline{)144}$

6) $14\overline{)812}$

7) $64\overline{)640}$

8) $94\overline{)188}$

9) $86\overline{)946}$

10) $47\overline{)423}$

11) $67\overline{)402}$

12) $47\overline{)141}$

13) $27\overline{)999}$

14) $92\overline{)644}$

15) $10\overline{)370}$

DAY 73

3-Digit by 2-Digit Division

1) 19) 8 5 5

2) 41) 3 6 9

3) 28) 3 6 4

4) 86) 6 8 8

5) 83) 6 6 4

6) 87) 7 8 3

7) 17) 9 6 9

8) 89) 1 7 8

9) 10) 6 1 0

10) 62) 2 4 8

11) 15) 4 2 0

12) 47) 6 5 8

13) 67) 2 0 1

14) 10) 2 6 0

15) 23) 5 2 9

DAY 74

3-Digit by 2-Digit Division

1) $19\overline{)760}$

2) $45\overline{)450}$

3) $96\overline{)480}$

4) $12\overline{)204}$

5) $11\overline{)990}$

6) $47\overline{)329}$

7) $51\overline{)663}$

8) $16\overline{)240}$

9) $63\overline{)189}$

10) $62\overline{)682}$

11) $17\overline{)323}$

12) $97\overline{)970}$

13) $81\overline{)891}$

14) $30\overline{)150}$

15) $20\overline{)220}$

DAY 75

3-Digit by 2-Digit Division

1) 15) 3 4 5

2) 11) 3 7 4

3) 45) 6 3 0

4) 12) 5 7 6

5) 11) 2 4 2

6) 16) 8 4 8

7) 37) 1 1 1

8) 14) 1 4 0

9) 29) 4 0 6

10) 16) 5 4 4

11) 83) 5 8 1

12) 19) 6 8 4

13) 54) 2 7 0

14) 41) 1 2 3

15) 71) 9 9 4

DAY 76

3-Digit by 2-Digit Division

1) 2 8) 7 0 0

2) 6 4) 1 2 8

3) 9 4) 2 8 2

4) 2 7) 6 2 1

5) 2 9) 1 1 6

6) 7 4) 2 2 2

7) 2 3) 7 8 2

8) 1 0) 7 9 0

9) 2 9) 9 8 6

10) 2 2) 3 7 4

11) 5 1) 2 5 5

12) 1 0) 8 8 0

13) 6 9) 1 3 8

14) 2 0) 2 6 0

15) 2 7) 6 2 1

TEENAGERS TEXTING IN THAILAND WILL SEND THE DIGIT 555 TO EACH OTHER TO INDICATE SOMETHING IS FUNNY.

In the Thai language, 5 is pronounced as 'ha', so 555 translated becomes ha-ha-ha! Want to hear a joke? Why was the math book sad? Because it had too many problems!

1)   4 5) 5 7 7   R

2)   4 6) 5 6 7   R

3)   2 8) 3 0 8   R

4)   4 1) 9 3 9   R

5)   3 4) 8 0 5   R

6)   3 1) 1 7 2   R

7)   3 3) 5 0 2   R

8)   1 1) 5 2 4   R

9)   3 4) 1 0 8   R

10)   5 0) 3 9 5   R

11)   1 7) 5 4 9   R

12)   4 9) 3 7 5   R

DAY 78

3-Digit by 2-Digit Division

1)

$2\ 0 \overline{)8\ 4\ 4}$ R

2)

$1\ 6 \overline{)1\ 2\ 9}$ R

3)

$2\ 4 \overline{)7\ 4\ 4}$ R

4)

$2\ 6 \overline{)7\ 2\ 0}$ R

5)

$3\ 7 \overline{)2\ 1\ 1}$ R

6)

$4\ 0 \overline{)4\ 0\ 8}$ R

7)

$4\ 4 \overline{)6\ 7\ 1}$ R

8)

$1\ 0 \overline{)5\ 0\ 0}$ R

9)

$2\ 3 \overline{)2\ 1\ 6}$ R

10)

$2\ 4 \overline{)3\ 6\ 6}$ R

11)

$3\ 0 \overline{)6\ 0\ 1}$ R

12)

$1\ 3 \overline{)6\ 2\ 2}$ R

DAY 79

3-Digit by 2-Digit Division

1) 2 0) 6 7 7   R

2) 4 3) 4 2 9   R

3) 1 8) 4 4 2   R

4) 1 0) 1 1 9   R

5) 5 0) 3 8 9   R

6) 3 7) 1 7 3   R

7) 4 1) 6 3 3   R

8) 3 2) 1 4 2   R

9) 2 8) 4 3 9   R

10) 4 5) 6 7 0   R

11) 1 5) 4 5 4   R

12) 1 9) 4 1 1   R

DAY 80

3-Digit by 2-Digit Division

1) 2 2) 4 4 2 R

2) 4 8) 8 8 3 R

3) 4 4) 2 7 0 R

4) 2 7) 1 4 7 R

5) 1 0) 5 5 3 R

6) 4 2) 2 6 0 R

7) 3 1) 6 2 7 R

8) 2 2) 4 4 7 R

9) 2 8) 1 0 4 R

10) 3 7) 4 3 2 R

11) 2 2) 9 6 2 R

12) 3 4) 6 3 0 R

DAY 81

3-Digit by 2-Digit Division

1) 1 2) 1 9 4   R

2) 2 7) 4 5 5   R

3) 3 8) 9 9 8   R

4) 4 9) 5 2 8   R

5) 3 2) 6 6 1   R

6) 1 3) 6 6 1   R

7) 4 3) 2 2 5   R

8) 1 0) 5 4 9   R

9) 3 7) 7 9 5   R

10) 2 1) 8 9 9   R

11) 3 3) 1 6 8   R

12) 2 9) 4 8 7   R

DAY 82 🐾
3-Digit by 2-Digit Division

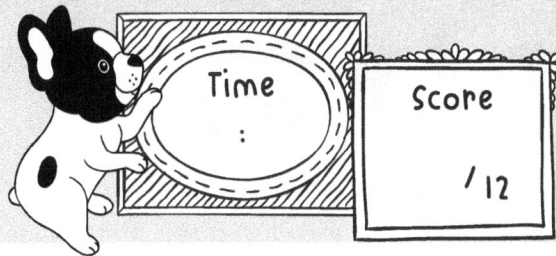

1) 13)321 R

2) 45)620 R

3) 29)502 R

4) 34)152 R

5) 20)793 R

6) 26)842 R

7) 30)937 R

8) 45)608 R

9) 12)474 R

10) 23)187 R

11) 30)513 R

12) 30)226 R

DAY 83

3-Digit by 2-Digit Division

1) 1 1) 3 1 8 R

2) 2 3) 9 7 4 R

3) 4 9) 4 3 7 R

4) 3 4) 5 5 1 R

5) 4 3) 2 3 8 R

6) 1 6) 9 4 2 R

7) 1 2) 8 3 1 R

8) 1 7) 2 3 1 R

9) 4 6) 5 7 1 R

10) 2 3) 4 7 6 R

11) 1 6) 6 8 5 R

12) 3 5) 5 0 1 R

DAY 84

3-Digit by 2-Digit Division

1) 2 1) 8 5 0 R

2) 3 9) 1 8 2 R

3) 3 7) 1 4 0 R

4) 2 5) 9 7 5 R

5) 4 1) 4 4 8 R

6) 1 7) 1 2 1 R

7) 1 5) 3 1 9 R

8) 2 1) 6 3 8 R

9) 2 9) 4 2 2 R

10) 4 4) 4 0 2 R

11) 1 4) 3 1 6 R

12) 3 7) 3 3 8 R

DIVIDING BY A LARGE NUMBER IN THE THOUSANDS? HERE'S A TRICK THAT MIGHT HELP.

If you need to divide by 1000, you can move the decimal point to the left once for each 0. For example, if I have 20000 ÷ 1000, I can move the decimal point left once to get 2000.0 twice to get 200.00 and a third time to get 20.000. The answer is 20! Pretty cool right? It made doing division with bigger numbers a lot easier for me!

DAY 85

4-Digit by 2-Digit Division

1) 28) 9 0 1 6

2) 17) 4 9 1 3

3) 13) 5 8 6 3

4) 11) 1 0 6 7

5) 43) 3 8 2 7

6) 61) 2 4 4 0

7) 10) 7 6 9 0

8) 43) 5 2 8 9

9) 47) 9 5 4 1

10) 29) 5 0 1 7

11) 32) 5 0 5 6

12) 10) 2 1 1 0

DAY 86

4-Digit by 2-Digit Division

1)

2 0) 3 4 8 0

2)

1 9) 2 6 2 2

3)

2 0) 9 7 4 0

4)

9 7) 9 9 9 1

5)

2 1) 5 8 1 7

6)

6 5) 1 6 9 0

7)

1 1) 7 7 1 1

8)

1 4) 3 8 9 2

9)

1 8) 6 4 4 4

10)

6 5) 5 2 0 0

11)

1 4) 5 6 5 6

12)

1 2) 1 0 5 6

DAY 87

4-Digit by 2-Digit Division

1) 35)1715

2) 10)8510

3) 50)5550

4) 49)4606

5) 14)8022

6) 19)4313

7) 12)5664

8) 31)6727

9) 47)1504

10) 52)4784

11) 11)1529

12) 41)6437

DAY 88

4-Digit by 2-Digit Division

1) 65)3835

2) 23)2553

3) 26)8658

4) 13)5967

5) 18)4698

6) 29)7627

7) 27)5778

8) 83)1162

9) 29)7105

10) 20)4820

11) 49)7889

12) 13)2119

DAY 89

4-Digit by 2-Digit Division

1) 18) 4 5 7 2

2) 12) 6 9 9 6

3) 12) 8 5 0 8

4) 10) 4 0 1 0

5) 13) 3 2 6 3

6) 15) 2 1 7 5

7) 15) 5 3 8 5

8) 14) 3 5 2 8

9) 16) 2 3 3 6

10) 17) 9 5 7 1

11) 25) 7 6 2 5

12) 75) 5 9 2 5

DAY 90 🐾

4-Digit by 2-Digit Division

1) 4 3) 3 3 9 7

2) 7 3) 7 8 8 4

3) 6 6) 9 7 6 8

4) 5 9) 8 7 3 2

5) 1 7) 6 2 7 3

6) 1 9) 6 5 9 3

7) 2 5) 5 0 7 5

8) 1 0) 7 5 7 0

9) 1 1) 7 2 7 1

10) 3 1) 6 5 4 1

11) 1 7) 5 1 3 4

12) 2 5) 7 0 7 5

DAY 91

4-Digit by 2-Digit Division

Time

:

Score

/ 12

1) 37)7437

2) 42)9030

3) 15)3300

4) 61)6161

5) 19)7619

6) 24)8136

7) 19)1387

8) 67)4891

9) 14)5754

10) 42)6342

11) 19)4313

12) 23)9844

DAY 92

4-Digit by 2-Digit Division

Time

:

1) 47)6157

2) 37)5291

3) 57)1083

4) 10)9970

5) 15)9915

6) 53)5353

7) 62)9486

8) 68)6392

9) 13)7319

10) 17)1445

11) 23)8809

12) 31)3193

I'M SURE YOU HAVE SEEN A RUBIK'S CUBE BEFORE, BUT DO YOU KNOW HOW MANY DIFFERENT WAYS THERE ARE TO SCRAMBLE ONE?

43,252,003,274,489,856,000! Wow, that's a big number. Like, really, really big! I can't comprehend that number, not in bananas, and I would certainly be asleep before I counted that many sheep! The Rubik's Cube was originally debuted internationally in 1980, and has become one of if not the best selling toy of all time. I tried to learn how to solve a Rubik's Cube once, and over an hour later I was still no closer to solving it than when I started!

DAY 93

4-Digit by 2-Digit Division

1)

3 5) 2 2 7 0   R

2)

1 3) 7 7 3 4   R

3)

3 1) 3 1 5 4   R

4)

5 8) 7 1 3 0   R

5)

4 6) 6 6 6 1   R

6)

8 6) 1 6 5 1   R

7)

2 2) 3 4 5 8   R

8)

8 3) 5 0 3 7   R

9)

9 9) 7 9 0 3   R

DAY 94

4-Digit by 2-Digit Division

1)

7 3) 3 7 7 1   R

2)

9 0) 6 4 5 4   R

3)

8 7) 1 8 2 7   R

4)

2 2) 8 8 8 9   R

5)

4 9) 7 7 5 7   R

6)

1 1) 2 1 7 1   R

7)

6 2) 6 1 7 2   R

8)

6 7) 8 4 2 6   R

9)

4 2) 5 3 9 0   R

DAY 95

4-Digit by 2-Digit Division

1)

3 6) 8 4 2 7   R

2)

5 7) 6 5 7 6   R

3)

1 8) 9 7 0 0   R

4)

2 5) 6 7 0 4   R

5)

8 1) 1 7 0 9   R

6)

4 2) 2 8 6 6   R

7)

2 1) 3 7 4 8   R

8)

2 9) 6 7 5 6   R

9)

5 4) 9 2 8 1   R

DAY 96

4-Digit by 2-Digit Division

Time
:

Score

/ 9

1) 5 1) 7 1 1 8   R

2) 6 7) 4 3 2 3   R

3) 7 9) 7 0 9 0   R

4) 9 7) 7 5 8 7   R

5) 8 1) 8 3 3 3   R

6) 2 9) 1 3 2 0   R

7) 1 4) 1 6 7 5   R

8) 3 4) 3 2 3 2   R

9) 8 3) 6 8 4 0   R

DAY 97

4-Digit by 2-Digit Division

1)

3 4) 4 8 1 2   R

2)

5 4) 4 6 2 7   R

3)

9 4) 8 1 8 3   R

4)

6 3) 8 2 0 0   R

5)

3 6) 3 7 8 7   R

6)

8 8) 3 5 2 8   R

7)

9 4) 3 7 1 8   R

8)

1 4) 1 9 1 6   R

9)

9 7) 4 8 4 8   R

DAY 98

4-Digit by 2-Digit Division

1)

7 2) 1 0 6 7   R

2)

7 0) 2 2 4 7   R

3)

1 3) 3 0 3 3   R

4)

3 3) 4 1 3 3   R

5)

7 3) 4 7 7 0   R

6)

6 4) 3 4 9 9   R

7)

8 8) 4 8 5 3   R

8)

8 1) 5 3 4 0   R

9)

4 5) 8 3 1 4   R

DAY 99

4-Digit by 2-Digit Division

Time

:

Score

/ 9

1)

$48 \overline{)1473}$ R

2)

$52 \overline{)7323}$ R

3)

$66 \overline{)5757}$ R

4)

$77 \overline{)9079}$ R

5)

$61 \overline{)8469}$ R

6)

$25 \overline{)4516}$ R

7)

$36 \overline{)1514}$ R

8)

$44 \overline{)7780}$ R

9)

$65 \overline{)3603}$ R

DAY 100

4-Digit by 2-Digit Division

1) 9 0) 7 7 9 4   R

2) 6 2) 2 6 2 9   R

3) 2 8) 6 1 7 0   R

4) 1 8) 5 7 9 0   R

5) 5 1) 4 6 3 0   R

6) 8 4) 3 2 8 8   R

7) 4 6) 9 3 7 9   R

8) 7 8) 9 4 1 8   R

9) 9 2) 4 2 7 7   R

FREE BONUS

I hope you had a blast learning about division with me.

Join me once again and dive into the captivating stories of extraordinary sport heroes and fearless entrepreneurs. I can't wait to share their remarkable tales of innovation and determination with you. In addition to the inspiring stories, I have included some fantastic coloring pages that will spark your creativity too!

So, what are you waiting for? Claim the freebies by scanning the QR code below or type riccagarden.com/ronny_freebies into your web browser.

Your Frenchie,
RONNY

(Note: You must be 16 years or older to sign up, so grab your parent for help if you need to.)

THE
REMARKABLE
STORIES
OF DREAMERS, SPORTS STARS AND CHILD HEROES

GET INSPIRED WITH

RONNY the FRENCHIE

THE Stars

YOU WILL SHINE
AMONG THEM LIKE STARS
IN THE SKY AS YOU HOLD FIRM
THE WORD OF LIFE Philippians

IN PEACE
I WILL LIE DOWN & SLEEP,
FOR YOU ALONE, O LORD,
MAKE ME DWELL IN
SAFETY.

Psalm 4:8

ANSWER KEY 🐾

DAY 1
(1) 3 (2) 12 (3) 11 (4) 13
(5) 11 (6) 10 (7) 2 (8) 15
(9) 2 (10) 13 (11) 13
(12) 24 (13) 13 (14) 14
(15) 25

DAY 2
(1) 7 (2) 43 (3) 11 (4) 29
(5) 7 (6) 19 (7) 16 (8) 4
(9) 31 (10) 14 (11) 11
(12) 14 (13) 10 (14) 5
(15) 13

DAY 3
(1) 2 (2) 31 (3) 10 (4) 4
(5) 5 (6) 29 (7) 31 (8) 22
(9) 43 (10) 6(11) 38
(12) 29 (13) 17 (14) 40
(15) 13

DAY 4
(1) 39 (2) 42 (3) 47
(4) 31 (5) 7 (6) 37 (7) 43
(8) 6 (9) 7 (10) 47 (11) 7
(12) 19 (13) 43 (14) 2
(15) 2

DAY 5
(1) 17 (2) 37 (3) 19
(4) 33 (5) 14 (6) 16
(7) 10 (8) 5 (9) 3 (10) 34
(11) 7 (12) 4 (13) 3
(14) 22 (15) 20

DAY 6
(1) 29 (2)19 (3) 9 (4)29
(5) 10 (6) 12 (7) 13
(8) 21 (9) 4 (10) 5 (11) 7
(12) 5 (13) 24 (14) 5
(15) 22

DAY 7
(1) 7 (2) 11 (3) 3 (4) 37
(5) 31 (6) 8 (7) 13 (8) 23
(9) 7 (10) 17 (11) 25
(12) 8 (13) 23 (14) 5
(15) 29

DAY 8
(1) 10 (2) 12 (3) 8 (4) 17
(5) 36 (6) 30 (7) 14 (8) 5
(9) 23 (10) 27 (11) 29
(12) 27 (13) 7 (14) 6
(15) 18

DAY 9
(1) 13 (2) 31 (3) 5 (4) 12
(5) 5 (6) 13 (7) 41 (8) 37
(9)11 (10) 43 (11) 33
(12) 17 (13) 41 (14) 9
(15) 7

DAY 10
(1) 19 (2) 41 (3) 21
(4) 31 (5) 36 (6) 31 (7) 4
(8) 31 (9) 19 (10) 11
(11) 25 (12) 10 (13) 12
(14) 8 (15) 9

DAY 11
(1)1 R 6 (2)31 R 1
(3)7 R 0 (4)10 R 7
(5)23 R 2 (6)15 R 3
(7)11 R 2 (8)10 R 2
(9)13 R 4 (10)19 R 1
(11)9 R 5 (12)1 R 2
(13)14 R 0 (14)7 R 2
(15)6 R 5

DAY 12
(1)15 R 4 (2)1 R 5
(3)8 R 5 (4)8 R 6
(5)41 R 0 (6)12 R 2
(7)2 R 3 (8)20 R 1
(9)41R 1 (10)8 R 0
(11)2 R 0 (12)3 R 3
(13)7 R 2 (14)11 R 6
(15)9 R 3

DAY 13
(1)30 R 1 (2)7 R 0
(3)12 R 1 (4)15 R 2
(5)18 R 3 (6)9 R 1
(7)31 R 1 (8)14 R 1
(9)8 R 1 (10)11 R 0
(11)5 R 0 (12) 11 R 0
(13)6 R 4 (14)9 R 3
(15)5 R 2

DAY 14
(1)13 R 4 (2)6 R 1
(3)18 R 1 (4)9 R 0
(5)11 R 0 (6)34 R 1
(7)8 R 1 (8)12 R 0
(9)24 R 0 (10)22 R 0
(11)11 R 6 (12)8 R 6
(13)21 R 1 (14)9 R 4
(15)15 R 1

DAY 15
(1)23 R 1 (2)13 R 0
(3)22 R 3 (4)3 R 6
(5)23 R 0 (6)9 R 3
(7)1 R 2 (8)5 R 0
(9)26 R 1 (10)4 R 2
(11)6 R 7 (12)4 R 3
(13)9 R 1 (14)11 R 5
(15)3 R 5

DAY 16
(1)2 R 3 (2)7 R 3
(3)9 R 8 (4)10 R 0
(5)4 R 1 (6)10 R 0
(7)16 R 2 (8)28 R 2
(9)15 R 4 (10)15 R 0
(11)22 R 2 (12)13 R 1
(13)2 R 2 (14)14 R 3
(15)17 R 0

DAY 17
(1)13 R 1 (2)3 R 0
(3)8 R 0 (4)7 R 1
(5)17 R 2 (6)8 R 0
(7)21 R 1 (8)1 R 3
(9)11 R 3 (10)25 R 2
(11)12 R 3 (12)2 R 3
(13)25 R 2 (14)6 R 1
(15)16 R 0

DAY 18
(1)13 R 2 (2)15 R 5
(3)19 R 1 (4)22 R 1
(5)10 R 0 (6)3 R 3
(7)8 R 6 (8)17 R 0
(9)11 R 7 (10)7 R 1
(11)8 R 2 (12)37 R 1
(13)6 R 6 (14)3 R 1
(15)24 R 2

DAY 19
(1)13 R 6 (2)15 R 1
(3)13 R 3 (4)12 R 4
(5)16 R 4 (6)11 R 1
(7)10 R 5 (8)10 R 1
(9)10 R 0 (10)24 R 2
(11)9 R 1 (12)31 R 1
(13)7 R 5 (14)5 R 4
(15)23 R 1

DAY 20
(1)2 R 7 (2)7 R 5
(3)19 R 1 (4)12 R 1
(5)9 R 6 (6)10 R 1
(7)4 R 5 (8)6 R 6
(9)11 R 1 (10)5 R 3
(11)3 R 4 (12)26 R 2
(13)17 R 1 (14)10 R 1
(15)12 R 2

DAY 21
(1) 154 (2) 53 (3) 41
(4) 26 (5) 43 (6) 199
(7) 109 (8)132 (9) 53
(10) 133 (11) 255
(12) 227 (13) 141
(14) 128 (15) 141

DAY 22
(1) 197 (2) 185 (3) 369
(4) 362 (5) 93 (6) 159
(7) 167 (8) 239 (9) 193
(10) 74 (11) 38 (12) 118
(13) 232 (14) 91
(15) 182

DAY 23
(1) 107 (2) 267 (3) 191
(4) 31 (5) 76 (6) 447
(7) 53 (8) 86 (9) 199
(10) 471 (11) 206
(12) 113 (13) 28
(14) 448 (15) 195

DAY 24
(1) 333 (2) 136 (3) 283
(4) 197 (5) 211 (6) 113
(7) 26 (8) 409 (9) 142
(10) 115 (11) 97
(12) 137 (13) 89
(14) 253 (15) 158

DAY 25
(1) 65 (2) 474 (3) 111
(4) 83 (5) 283 (6) 223
(7) 288 (8) 69 (9) 130
(10) 148(11) 94 (12) 92
(13)226 (14) 137
(15) 152

DAY 26
(1) 73 (2) 151 (3) 407
(4) 131 (5) 76 (6) 71
(7) 394 (8) 154
(9) 241(10) 257 (11)
125 (12) 278 (13) 98
(14) 179 (15) 283

DAY 27
(1) 389 (2) 221 (3) 130
(4) 148 (5) 171 (6) 104
(7) 109 (8) 123 (9) 145
(10) 211 (11) 113
(12) 110 (13) 317
(14) 85 (15) 367

DAY 28
(1) 101 (2) 122 (3) 215
(4) 281 (5) 77 (6) 44
(7) 228 (8) 79 (9) 491
(10) 57 (11) 161
(12) 281 (13) 42
(14) 107 (15) 107

DAY 29
(1) 181 (2) 408 (3) 132
(4) 158 (5) 131 (6) 79
(7) 499 (8) 167 (9) 71
(10) 302 (11) 201
(12) 188 (13) 475
(14) 62 (15) 444

DAY 30
(1)391 (2) 206 (3) 41
(4)491 (5) 89 (6) 49
(7) 445 (8) 218
(9) 29(10) 214 (11) 361
(12)109 (13) 111
(14) 218 (15) 347

DAY 31
(1)96 R 0 (2)39 R 0
(3)237 R 1 (4)105 R 2
(5)72 R 1 (6)39 R 0
(7)166 R 1 (8)44 R 2
(9)95 R 1 (10)151 R 0
(11)206 R 1 (12) 35 R 1

DAY 32
(1)52 R 1 (2)13 R 3
(3)24 R 0 (4)108 R 7
(5)105 R 1 (6)36 R 3
(7)42 R 3 (8)41 R 3
(9)68 R 3 (10)26 R 5
(11)20 R 2 (12)102 R 3

DAY 33
(1)131 R 6 (2)16 R 6
(3)78 R 1 (4)77 R 3
(5)36 R 1 (6)128 R 2
(7)178 R 1 (8)84 R 7
(9)64 R 1 (10)67 R 0
(11)157 R 3 (12)103 R 0

DAY 34
(1)129 R 1 (2)55 R 0
(3)82 R 5 (4)276 R 0
(5)152 R 4 (6)152 R 0
(7)400 R 0 (8)178 R 2
(9)39 R 2 (10)79 R 5
(11)140 R 2 (12)18 R 4

DAY 35
(1)40 R 0 (2)464 R 0
(3) 24 R 4 (4)62 R 2
(5)140 R 0 (6)54 R 2
(7)185 R 2 (8)121 R 4
(9)192 R 3 (10)69 R 1
(11)17 R 6 (12)142 R 3

ANSWER KEY 🐾

DAY 36
(1)103 R 3 (2)108 R 1
(3)147 R 4 (4)69 R 3
(5)47 R 3 (6)96 R 4
(7)102 R 0 (8)125 R 0
(9)78 R 4 (10)173 R 1
(11)46 R 2 (12)29 R 2

DAY 37
(1)29 R 2 (2)11 R 2
(3)302 R 2 (4)230 R 2
(5)61 R 1 (6)179 R 0
(7)82 R 1 (8)14 R 6
(9)141 R 6 (10)31 R 2
(11)252 R 2 (12)32 R 1

DAY 38
(1)50 R 2 (2)475 R 0
(3)80 R 4 (4)164 R 4
(5)195 R 3 (6)26 R 3
(7)50 R 1 (8)87 R 4
(9)53 R 1 (10)73 R 0
(11)424 R 0 (12)123 R 0

DAY 39
(1)150 R 4 (2)23 R 2
(3)380 R 1 (4)19 R 2
(5)19 R 1 (6)406 R 0
(7)25 R 1 (8)111 R 2
(9)164 R 2 (10)197 R 0
(11)79 R 0 (12)105 R 6

DAY 40
(1)107 R 6 (2)50 R 4
(3)131 R 2 (4)35 R 1
(5)48 R 8 (6)271 R 0
(7)59 R 5 (8)55 R 2
(9)24 R 4 (10)39 R 0
(11)70 R 0 (12)139 R 3

DAY 41
(1) 691 (2) 2131 (3) 153
(4) 4388 (5) 379
(6) 2419 (7) 1633
(8) 1765 (9) 963
(10) 3811 (11) 1768
(12) 1283

DAY 42
(1) 235 (2) 1534 (3) 653
(4) 791 (5) 2333 (6)2002
(7) 453 (8) 423 (9) 739
(10) 1458 (11) 2836
(12) 775

DAY 43
(1) 1394 (2) 751
(3) 3260 (4) 2399
(5) 3107 (6) 3829
(7) 4106 (8) 1709
(9) 4471 (10) 589
(11)195 (12) 1289

DAY 44
(1) 1188 (2) 1867
(3) 1116 (4) 853
(5) 1158 (6) 765 (7) 731
(8) 840 (9) 1543
(10) 1611 (11) 1486
(12) 433

DAY 45
(1) 317 (2) 575 (3) 1012
(4) 1334 (5) 1382
(6) 3779 (7) 1103
(8) 1751(9) 911
(10) 639 (11) 663
(12) 4597

DAY 46
(1) 426 (2) 4100
(3) 2563 (4) 1087
(5) 924 (6) 1581
(7) 1843 (8) 2279
(9) 294 (10) 1137
(11) 1462 (12) 1660

DAY 47
(1) 449 (2) 3634 (3) 967
(4) 191 (5) 1588
(6) 3781 (7) 221
(8) 1278 (9) 627
(10) 155 (11) 1109
(12) 710

DAY 48
(1)2267 (2)1705
(3) 4190(4) 1598
(5) 1649 (6)333
(7) 1969 (8) 1595
(9) 4611 (10) 509
(11) 4607 (12) 1343

DAY 49
1) 2361 (2) 2186
(3) 971 (4) 1421 (5) 931
(6) 3299 (7) 901
(8) 4651 (9) 3307
(10) 1180 (11) 1453
(12) 361

DAY 50
1) 4621 (2) 1027
(3) 1985 (4) 2309
(5) 1353 (6) 1884
(7) 1619 (8) 475 (9) 949
(10) 708 (11) 1859
(12) 511

DAY 51
(1)1389 R 4 (2)2213 R 0
(3)1838 R 3 (4)1522 R 1
(5)948 R 0 (6)1260 R 3
(7)266 R 2 (8)1056 R 7
(9)170 R 4 (10)1167 R 5
(11)2115 R 1
(12)821 R 6

DAY 52
(1)3077 R 1 (2)416 R 0
(3)1942 R 4 (4)586 R 7
(5)2511 R 0 (6)292 R 1
(7)1356 R 3 (8)633 R 3
(9)1620 R 4 (10)488 R 0
(11)3577 R 0
(12)1088 R 2

DAY 53
(1)576 R 5 (2)908 R 6
(3)3759 R 0 (4)1114 R 3
(5)968 R 2 (6)292 R 3
(7)756 R 1 (8)1187 R 0
(9)1630 R 0 (10)210 R 5
(11)1145 R 1
(12)1821 R 2

DAY 54
(1)1746 R 1 (2)3295 R 1
(3)545 R 2 (4)571 R 1
(5)1652 R 1 (6)243 R 7
(7)1149 R 1 (8)999 R 0
(9)1436 R 0
(10)1974 R 1
(11)373 R 2 (12)422 R 2

DAY 55
1)4579 R 0 (2)175 R 0
(3)464 R 1 (4)697 R 8
(5)774 R 6 (6)1329 R 0
(7)693 R 7 (8)252 R 0
(9)1748 R 3 (10)244 R 0
(11)1202 R 1
(12)1304 R 0

DAY 56
(1)826 R 5 (2)441 R 4
(3)1144 R 3 (4)1093 R 4
(5)665 R 4 (6)202 R 7
(7)3974 R 1 (8)718 R 0
(9)946 R 7 (10)621 R 1
(11)563 R 8 (12)835 R 4

DAY 57
(1)749 R 3 (2)3071 R 1
(3)935 R 3 (4)580 R 4
(5)2486 R 3 (6)1004 R 3
(7)1213 R 1 (8)739 R 3
(9)1182 R 1 (10)1591 R
5 (11)883 R 1 (12)2044
R 1

DAY 58
(1)528 R 3 (2)661 R 7
(3)1856 R 1 (4)3672 R 0
(5)555 R 0 (6)1480 R 4
(7)1396 R 4 (8)2264 R 3
(9)1248 R 2 (10)1085 R
4 (11)923 R 4 (12)588 R
4

DAY 59
(1)1701 R 1 (2)3147 R 0
(3)1723 R3 (4)676 R 6
(5)368 R 1 (6)3211 R 0
(7)1280 R 0 (8)649 R 2
(9)530 R 4 (10)1278 R 5
(11)637 R 3 (12)2145 R
1

DAY 60
(1)1209 R 3 (2)1583 R 3
(3)1590 R 4 (4)563 R 2
(5)1492 R 1 (6)999 R 2
(7)539 R 5 (8)980 R 6
(9)287 R 7 (10)685 R 3
(11)293 R 4 (12)564 R 1

DAY 61
(1)19151 R 2 (2)3173 R
7 (3)18138 R 1(4)10924
R 5 (5)2577 R2
(6)15400 R 3 (7)1922 R
5 (8)15505 R 1

DAY 62
(1)4178 R 3 (2)7777 R 4
(3)7163 R 7 (4)5653 R 1
(5)11741 R 4 (6)18769
R 1 (7)5625 R 7
(8)26857 R 0

DAY 63
(1)6916 R 2 (2)10557 R
0 (3)6066 R 2 (4)3756 R
4 (5)2596 R 1 (6)26444
R 1 (7)10247 R 0
(8)11754 R 5

DAY 64
(1)4822 R 0 (2)22211 R
2 (3)21487 R 1
(4)28998 R 1 (5)3752 R
1 (6)12544 R 0 (7)3568
R 1 (8)15451 R 2

DAY 65
(1)2481 R 5 (2)5506 R 4
(3)8669 R 2 (4)25255 R
0 (5)8179 R 4 (6)8973 R
4 (7)2890 R 3 (8)17685
R 1

DAY 66
(1)19886 R 0 (2)4848 R
2 (3)23543 R 2
(4)17376 R 2 (5)1738 R
4 (6)4593 R 5 (7)6395 R
2 (8)3269 R 2

DAY 67
(1)4778 R 3 (2)2972 R 3
(3)8103 R 3 (4)6340 R 1
(5)9544 R 2 (6)10001 R
2 (7)10930 R 6
(8)11934 R 2

DAY 68
(1)13445 R 2 (2)10160
R 5 (3)9051 R 7
(4)11487 R 2 (5)2505 R
5 (6)7708 R 3 (7)12098
R 1 (8)10521 R 2

DAY 69
(1) 4 (2) 53 (3) 6 (4) 8
(5) 7 (6) 8 (7) 9 (8) 5
(9) 10 (10) 7 (11) 25
(12) 15 (13) 13 (14) 5
(15) 13

DAY 70
(1) 10 (2) 15 (3) 18 (4) 5
(5) 5 (6) 18 (7) 33 (8) 7
(9) 3 (10) 7 (11) 23
(12) 5 (13) 12 (14) 7
(15) 4

ANSWER KEY 🐾

DAY 71
(1) 11 (2) 4 (3) 26 (4) 11
(5) 26 (6) 21 (7) 6 (8) 4
(9) 11 (10) 11 (11) 22
(12) 7 (13) 21 (14) 45
(15) 7

DAY 72
(1) 83 (2) 11 (3) 9 (4) 2
(5) 6 (6) 58 (7)10 (8) 2
(9) 11 (10) 9 (11) 6
(12) 3 (13) 37 (14) 7
(15) 37

DAY 73
(1) 45 (2) 9 (3) 13 (4) 8
(5) 8 (6) 9 (7) 57 (8) 2
(9) 61 (10) 4 (11) 28
(12) 14 (13) 3 (14) 26
(15) 23

DAY 74
(1) 40 (2) 10 (3) 5 (4) 17
(5) 90 (6)7 (7) 13 (8) 15
(9) 3 (10) 11 (11) 19
(12) 10 (13) 11 (14) 5
(15) 11

DAY 75
(1) 23 (2) 34 (3) 14
(4) 48 (5) 22 (6) 53 (7) 3
(8) 10 (9) 14 (10) 34
(11) 7 (12) 36 (13) 5
(14) 3 (15) 14

DAY 76
(1) 25 (2) 2 (3) 3 (4) 23
(5) 4 (6) 3 (7) 34 (8) 79
(9) 34 (10) 17 (11) 5
(12) 88 (13) 2 (14) 13
(15) 23

DAY 77
(1)12 R 37 (2)12 R 15
(3)11 R 0 (4)22 R 37
(5)23 R 23 (6)5 R 17
(7)15 R 7 (8)47 R 7 (9)3
R 6 (10)7 R 45 (11)32 R
5 (12)7 R 32

DAY 78
(1)42 R 4 (2)8 R 1 (3)31
R 0 (4)27 R 18 (5)5 R 26
(6)10 R 8 (7)15 R 11
(8)50 R 0 (9)9 R 9
(10)15 R 6 (11)20 R 1
(12)47 R 11

DAY 79
(1)33 R 17 (2)9 R 42
(3)24 R 10 (4)11 R 9
(5)7 R 39 (6)4 R 25
(7)15 R 18 (8)4 R 14
(9)15 R 19 (10)14 R 40
(11)30 R 4 (12)21 R 12

DAY 80
(1)20 R 2 (2)18 R 19
(3)6 R 6 (4)5 R 12 (5)55
R 3 (6)6 R 8 (7)20 R 7
(8)20 R 7 (9)3 R 20
(10)11 R 25 (11)43 R 16
(12)18 R 18

DAY 81
(1)16 R 2 (2)16 R 23
(3)26 R 10 (4)10 R 38
(5)20 R 21 (6)50 R 11
(7)5 R 10 (8)54 R 9
(9)21 R 18 (10)42 R 17
(11)5 R 3 (12)16 R 23

DAY 82
(1)24 R 9 (2)13 R 35
(3)17 R 9 (4)4 R 16
(5)39 R 13 (6)32 R 10
(7)31 R 7 (8)13 R 23
(9)39 R 6 (10)8 R 3
(11)17 R 3 (12)7 R 16

DAY 83
(1)28 R 10 (2)42 R 8
(3)8 R 45 (4)16 R 7
(5)5 R 23 (6)58 R 14
(7)69 R 3 (8)13 R 10
(9)12 R 19 (10)20 R 16
(11)42 R 13 (12)14 R 11

DAY 84
(1)40 R 10 (2)4 R 26
(3)3 R 29 (4)39 R 0
(5)10 R 38 (6)7 R 2
(7)21 R 4 (8)30 R 8
(9)14 R 16 (10)9 R 6
(11)22 R 8 (12)9 R 5

DAY 85
(1) 322 (2) 289 (3) 451
(4) 97 (5) 89 (6) 40
(7) 769 (8) 123 (9) 203
(10) 173 (11) 158
(12) 211

DAY 86
(1) 174 (2) 138 (3) 487
(4) 103 (5) 277(6) 26
(7) 701 (8) 278 (9) 358
(10) 80 (11) 404 (12) 88

DAY 87
(1) 49 (2) 851 (3) 111
(4) 94 (5) 573 (6) 227
(7) 472 (8) 217 (9) 32
(10) 92 (11) 139
(12) 157

DAY 88
(1) 59 (2) 111 (3) 333
(4) 459 (5) 261 (6) 263
(7) 214 (8) 14 (9) 245
(10) 241 (11) 161
(12) 163

DAY 89
(1) 254 (2) 583 (3) 709
(4) 401 (5) 251 (6) 145
(7) 359 (8) 252 (9) 146
(10) 563 (11) 305
(12) 79

DAY 90
(1) 79 (2) 108 (3) 148
(4) 148 (5) 369 (6) 347
(7) 203 (8) 757 (9) 661
(10) 211 (11) 302
(12)283

DAY 91
(1) 201 (2) 215 (3) 220
(4) 101 (5) 401 (6) 339
(7) 73 (8) 73 (9) 411
(10) 151 (11) 227
(12) 428

DAY 92
(1) 131 (2) 143 (3) 19
(4) 997 (5) 661 (6) 101
(7) 153 (8) 94 (9) 563
(10) 85 (11) 383
(12) 103

DAY 93
(1)64 R 30 (2)594 R 12
(3)101 R 23 (4)122 R 54
(5)144 R 37 (6)19 R 17
(7)157 R 4 (8)60 R 57
(9)79 R 82

DAY 94
(1)51 R 48 (2)71 R 64
(3)21 R 0 (4)404 R 1
(5)158 R15 (6)197 R 4
(7)99 R 34 (8)125 R 51
(9)128 R 14

DAY 95
(1)234 R 3 (2)115 R 21
(3)538 R 16 (4)268 R 4
(5)21 R 8 (6)68 R 10
(7)178 R 10 (8)232 R 28
(9)171 R 47

DAY 96
(1)139 R 29 (2)64 R 35
(3)89 R 59 (4)78 R 21
(5)102 R 71 (6)45 R 15
(7)119 R 9 (8)95 R 2
(9)82 R 34

DAY 97
(1)141 R 18 (2)85 R 37
(3)87 R 5 (4)130 R 10
(5)105 R 7 (6)40 R 8
(7)39 R 52 (8)136 R 12
(9)49 R 95

DAY 98
(1)14 R 59 (2)32 R 7
(3)233 R4 (4)125 R 8
(5)65 R 25 (6)54 R 43
(7)55 R 13 (8)65 R 75
(9)184 R 34

DAY 99
(1)30 R 33 (2)140 R 43
(3)87 R 15 (4)117 R 70
(5)138 R 51 (6)180 R 16
(7)42 R 2 (8)176 R 36
(9)55 R 28

DAY 100
(1)86 R 54 (2)42 R 25
(3)220 R 10 (4)321 R 12
(5)90 R 40 (6)39 R 12
(7)203 R 41 (8)120 R 58
(9)46 R 45